CYBERSECURITY AS A FISHING GAME

Developing Cybersecurity in the Form of Fishing Game and what Top Management should Understand

Dr. Tan Kian Hua (Author)
Dr. Vladimir Biruk (Coauthor)

PARTRIDGE

Copyright © 2021 by Dr. Tan Kian Hua, Dr. Vladimir Biruk.

ISBN:	Hardcover	978-1-5437-6523-6
	Softcover	978-1-5437-6524-3
	eBook	978-1-5437-6522-9

All rights reserved. No part of this book may be used or reproduced by any means, graphic, electronic, or mechanical, including photocopying, recording, taping or by any information storage retrieval system without the written permission of the author except in the case of brief quotations embodied in critical articles and reviews.

Because of the dynamic nature of the Internet, any web addresses or links contained in this book may have changed since publication and may no longer be valid. The views expressed in this work are solely those of the author and do not necessarily reflect the views of the publisher, and the publisher hereby disclaims any responsibility for them.

Print information available on the last page.

To order additional copies of this book, contact
Toll Free +65 3165 7531 (Singapore)
Toll Free +60 3 3099 4412 (Malaysia)
orders.singapore@partridgepublishing.com

www.partridgepublishing.com/singapore

Contents

Foreword vii
Acknowledgments xi
Introduction xv

Chapter 1 Information Security and the CIA Triad 1
Chapter 2 Cyberattack and the Fishing Rod 11
Chapter 3 The Hackers' Game and Strategies for
               Hacking 23
Chapter 4 Powering Up the Game: Upgrading
               to the Fishing Net 51
Chapter 5 The Current Game: Fishing Boat 66

Afterword 77
Final Thoughts 83
References 85
About the Authors 89
Index 95

Foreword

The idea to write a joint book arose in dissertation research on the Internet of Things and cybersecurity in cooperation with a young scientist, a graduate of American LIGS University, Dr. Tan Kian Hua, and professor of LIGS University, Dr. Vladimir Biruk.

Enthusiasm and modern knowledge of Tan Kian Hua, working at the peak of the information technology industry's development, were aimed at the Internet of Things (IoT) safe adaptation to facilitate people's lives and business operations. Partner and coauthor of the book, scientific adviser Dr. Vladimir Biruk went through a half-century evolutionary path developing and implementing information and communication technologies (ICT).

Professor Biruk began using the first electronic computers for production processes and work research in FORTRAN, Pascal, and others, having mastered the most modern approaches, both in software and transaction security. That was the basis of programming and machine data processing in the late sixties and within the seventies of the twentieth century.

Furthermore, with the emergence and development of the personal computer and the internet, computing technology was used for complex mathematical calculations and programming for scientific research in many industries. The range of business and life areas is summarized in the authors' biographies, demonstrating the breadth of their knowledge and skills.

There is a synergy of a modern specialist in a significantly developing industry and a professional who has passed the evolutionary path of information technology implementation. They provide an excellent basis for the discourse of both the historical stages in the development of ICT and the future digital world's principles within the postindustrial economy and human life in the future.

Despite the differences in experience and approaches, a modern scientist with a sense of risk appetite for cooperation and an expert who understands risk assessment rules, which allow to optimize approaches' adaptability, create a productive research team.

Moreover, the convergence of models for calculating risk and uncertainty with the level of modern science creativity makes it possible to arrive at an optimal understanding of the information and communication system's tendency and workers' safety within this system. The combination of two opposites led to high-quality preparation of the thesis and an excellent result. It allows us to convincingly offer to the professional community an integrated view of a scientific and practical product and its safe adaptation to changing systems.

Besides, extensive practice in communication with leading worldwide professional companies—such as IBM, Oracle, Sybase, AT&T, and many others—allows us to offer trusted models of ICT security and its prospective areas.

We hope that our perspective will help security professionals and ordinary users of information and communication systems. The reader will find available recipes to ensure their safety, maintain operational continuity, and restore business processes.

Acknowledgments

Dr. Tan Kian Hua

No word can express my extreme gratitude to my parents for their love, prayers, care, sacrifices for educating and preparing me for my future, and *for always being the people I could turn to during those dark and desperate years.* Also, I express my gratitude to both my elder and younger sisters for their love, understanding, prayers, and continuing support to complete this book. *You stood by me during every struggle and all my successes;* it was a great comfort and relief. Your encouragement when the times got rough is much appreciated and duly noted. My heartfelt thanks.

I would like to express my deep and sincere gratitude to Professor Vladimir Biruk for providing invaluable guidance throughout this book. His dynamism, vision, sincerity, and motivation have deeply inspired me. It was a great privilege and honor to work under his guidance. I am extremely grateful for what he has offered me. I would also like to thank him for his friendship, empathy, and great sense of humor. I am extending my heartfelt thanks to his wife and family for their acceptance and patience during the time I had with him.

I want to thank everyone who ever said anything positive to me or taught me something. I heard it all, and it meant something.

Finally, my thanks go to all the people who have supported me to complete the research work directly or indirectly. Thanks to everyone on my publishing team.

Dr. Professor Vladimir Biruk

Like most authors who have reached the final stage of their intellectual work, we are aware of the weight of everyone who contributed to this result and possible success after the publication of the book. Of course, each of us has made a significant contribution to the scientific research preceding the description in the book. Our colleagues and industry professionals have contributed to the discourses about the appropriateness and practicality of the publication.

I want to express our special gratitude to our families, who endured all the distractions of us from family affairs for many months and years. I also would like to express our deep appreciation for our teachers—professors Luksha L. K., Ailamazyan A. K., Nikolaev V. B.—colleagues, and students who contributed to forming our way of thinking to the formulation of concepts suitable for practitioners and beginners.

We recognize that our publication is only a tiny link in a large chain of cross-disciplinary knowledge that promotes ICT development. We are confident other professionals will use this information to improve the security system in such a rapidly developing ICT industry.

Introduction

Before starting an introduction to our book's subject, talking about how business leaders and ordinary users of modern ICT should respond to experienced and novice hackers' devastating attacks, we would like to remind the history and evolution of communications development.

It is well known that communication between people began when two persons, Adam and Eve, appeared. However, information transfer technologies appeared much later when the need arose for human communities to communicate and transfer information at a distance.

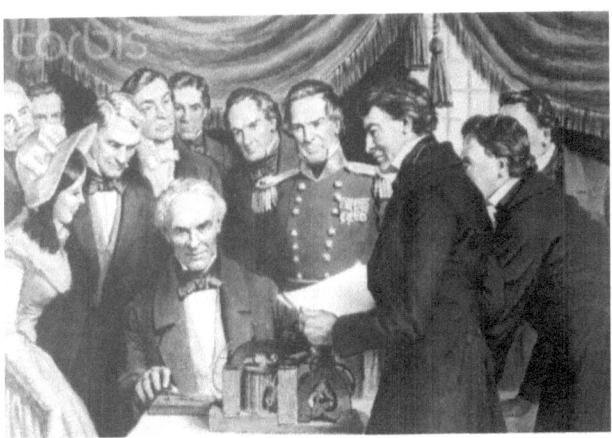

Primary technologies were noted 4,000 BC to transmit information using drums and the human voice using additional devices. The most crucial stage was the invention of writing systems in ancient China (Cangjie), Egypt, and Phoenicia (Alphabet). The rise of printed technologies became the foundation to information transmission, world religious extension, and statehood's legal systems exploration. That led to the specificity of information security paradigms formation under the control of the clergy and power elites.

The improvement of such methods took hundreds and thousands of years until the advent of more modern devices for transmitting information at a distance (H. Hertz, N. Tesla, A. Popov, G. Marconi, K. Brown, etc.). The most impressive development was the telegraph's invention (W. Cooke, C. Wheatstone, S. Morse). Here's the first telegraphic message in Morse code on May 24, 1844: "What hath God wrought!" The proliferation of telegraph, telephone, and radio communications, even across Europe and North America, was selective, taking into account manageability and the new paradigm of information security. The complexity and high cost of maintaining the infrastructure had significantly increased the cost of data transmission. For example, a telegraph message about Alaska's cession by the Russian Empire to the United States of America cost the sender $10,000.

Later on, the twentieth and twenty-first centuries, which accelerated the arrival of new industrial revolutions, contributed to the talented inventors of electronic computers and digital information and communication technologies, like Alan Turing and David Clarke. The use of new means of telecommunications in trade, economic relations, and the military sphere influenced the interstate balance of power globally. All this could affect the possibility and methods of receiving transmitted information, espionage, and surveillance. The peak of perfection of secret tracking and hacking of systems was reached during the Cold War.

Digital ICT in the new century is marked by rapid progress in the production of various information communication tools, such as IoT and IIoT, that are transforming almost all areas of human life. Among new communication technologies, 4G, 5G, audiovisual forms of information presentation take the top place. Trends in mobile devices and services and social networks are marked by an increase in new challenges and threats.

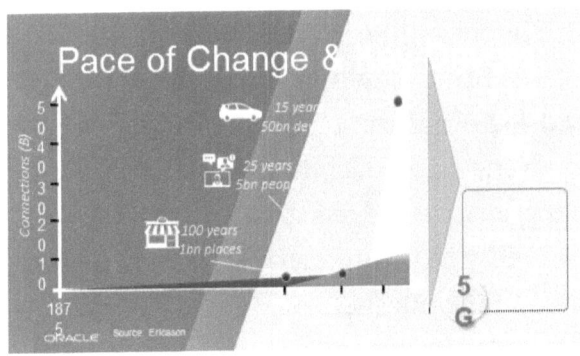

A paradigm shift is taking place toward ensuring international information security and legal regulation of digital ICT.

Our world is developing rapidly, and the internet now plays a crucial role in almost every individual's life. The increase in dependence on the internet has led to an increase in the risks of being victims of cyberattacks. Did you know that a hacker attack takes place every thirty-nine seconds? This information is based on a Clark School study carried out at the University of Maryland. One in three Americans falls victim to these hacks each year.

Averagely, the global cost of data breaches across SMBs is $3.9 million, and this does not affect just the finances and businesses but also their reputation. This cost is often more significant for public companies because many things are at stake.

It is interesting to know that the number of connected Internet of Things (IoT) devices around the world is about thirty-one billion in 2020. Will this figure increase to about seventy-five billion IoT devices by 2025? There is a need for organizations to carry out fundamental changes in their approach to cybersecurity. Companies need to reprioritize budgets to ensure that they align with our new reality in this modern world where everyone is connected to the internet.

In the United States, over five hundred thousand cybersecurity jobs are unfilled; and over the past five years, postings increased by 74 percent. In fact, the number of unfilled cybersecurity jobs around the world is already more than four million. The highest-paid cybersecurity positions are cybersecurity engineers. It is expected that the amount of money that organizations will spend globally on cybersecurity by 2021 will be approximately $6 trillion.

Cybersecurity as a Fishing Game is the world's first cybersecurity book for all cybersecurity professionals or those who want to understand cybersecurity in terms of a fishing game. It will walk you through the way how cybersecurity evolved and educate you about the hacking game.

Regardless of the innovations and methods used in the cyber game, what will be the common ground? In the cyber game, there will also be rules and principles that the reader needs to know. This book is primarily focused on exposing these rules and principles guiding the activities of cyber hackers. And by understanding them, professionals can review how to enhance their defensive game. Are you ready for this journey? Enjoy your reading!

Chapter 1
Information Security and the CIA Triad

With the increasing reach of www (world wide web), today's internet connection has enabled businesses of all scales and from any location to expand into new and larger markets. Cyberspace has also provided opportunities for businesses to engage more productively and work more efficiently. With such an enabler as the internet, companies can now adopt cloud computing to reduce costs and streamline operations. They now use email and maintain a website, which have further helped to open more doors to businesses.

But with the creation of opportunities in the digital information age, there is also a significant increase in the rate of theft of digital information, fraud, and physical theft. As earlier mentioned, a hacker attack occurs every thirty-nine seconds, and data breaches in different locations worldwide cost SMBs about $3.9 million. This affects both the reputations of businesses as well as their finances.

Therefore, every business with access to the internet must be prepared to create information security that will enhance business and consumer confidence. The core of cybersecurity is information security; the requirement will be for safe utilization, flow, and storage of information. To understand these fundamental concepts will be to understand the CIA triad.

CIA

The foundation of information security makes up the CIA triad's core objective, which is to maintain and ensure that the three elements of the information technology system and business data are effectively in place.

CIA triad in the cybersecurity context refers to *confidentiality, integrity,* and *availability.* It is a model commonly used to design and guide policies for information security within any corporation or enterprise. It can be commonly referred to as the AIC triad (availability, integrity, and confidentiality). Regardless of the naming arrangement, these three essential elements are considered to be the most crucial components of information security.

What is the importance of the CIA triad? Those familiar with the basics of cybersecurity will undoubtedly understand the importance of the three concepts. But why think of them as a triad of connected ideas instead of considering them separately? In the face of the increasing array of security services, software, and techniques in the marketplace, the CIA triad helps you make sense of everything.

Instead of just throwing money around on vague problems plaguing your organization, it is better to ask focused questions regarding the best way to spend money to enhance your organization's cybersecurity. The CIA triad helps you to ask salient questions like the following:

- *Confidentiality.* Does this tool I want to purchase make my organization's information more secure? When in place, the triad will ensure that sensitive information in any organization will only be disclosed to authorized parties.
- *Availability.* Will our data still be readily available to the individuals who need it after beefing up our infrastructure? The CIA triad helps to guarantee that the data will only be accessible to authorized parties when requested.
- *Integrity.* Will this service help secure the integrity of our data? The data will be prevented from unauthorized modification.

Another benefit of arranging these three concepts in a triad is making it quite clear to us that they exist, and they are usually in tension with one another in most cases. CIA triad offers us a basic understanding of cybersecurity, so let us examine each CIA triad component in detail.

Confidentiality

To understand the concept of confidentiality, we can think of it as roughly equivalent to privacy. The measures for ensuring confidentiality will be the design to prevent sensitive information from reaching the wrong party. This will also ensure that the right party authorized to use the information will be able to access it.

A good number of information systems accommodate data that have some level of sensitivity. This could be proprietary business information that would benefit competitors should they lay their hands on it. It could also be personal information regarding the customers, clients, or employees of an organization. Since confidential information is extremely valuable, systems are usually under frequent attack as cyber hackers are trying to exploit identified vulnerabilities.

When an organization fails to maintain confidentiality, it implies that someone else who should not access their data has managed to lay hands on private information. This could actually be by accident or through the intentional behavior of cyber hackers.

The weakest link in the cybersecurity infrastructure will be the people. Due to the increasing number of cyberattack cases, staff training is essential to ensure confidentiality, safeguarding data, and protecting the weakest link—employees. Conducting training will help familiarize all employees in the organization with the best practices in terms of cybersecurity and authorize personnel to know the risk factors and guard against them.

Training can stretch all the way to include strong passwords and password-related best practices and guard against social engineering methods. This is to prevent employees from bending data-handling rules with good intentions and resulting in potentially disastrous results.

If the access to information falls into the wrong hand, the damage can be unbearable for the organization. The damage will be categorized according to the data based on the amount or type of damage that could be done on the data should it fall into unintended hands. The level of control, be it more or less stringent to be implemented, will depend on the potential damage categories.

Majority of what people consider to be cybersecurity, which includes things that restrict access to data, fall within the confidentiality category. It includes authentication, which has to do with all the processes that allow any system to ascertain if truly a particular user is who they claim to be. It involves passwords as well as other techniques that help establish an identity, like biometric verification.

Data encryption is a standard method of ensuring confidentiality. The market norm of practice will be two-factor authentication in addition to user IDs and passwords. Other options will include biometric verification, soft tokens, cryptographic keys, and security tokens. Taking a more precautious method will be for extremely sensitive documents. They can be stored only on air-gapped computers, disconnected storage devices, or hard copies for highly sensitive information only. The organization must also minimize the number of places where the information will appear and reduce the number of times information is actually transmitted to complete a required transaction.

> Confidentiality is the heart of the relationship
> of trust between a subject and an object.

The second example of confidentiality is authorization, and this helps to ascertain who specifically has the right to access data. A system can detect who you are, but that does not guarantee that it will allow you to access all its data. Creating need-to-know mechanisms for data access is one of the most crucial ways of enforcing confidentiality. It ensures that accounts of users that have been hacked or those who have gone rogue cannot in any way compromise sensitive data. One of the most common ways operating systems enforce confidentiality is by having several files that can only be accessed by an admin or someone who created it.

Integrity

This aspect of the triad has to do with protecting information from unauthorized alteration, which provides a measure of assurance in the accuracy and completeness of data. The data that needs to be protected in this case will be data at rest (stored on systems) and data in transit (transmitted between point A to point B). Those controlling data at this level need to ensure that there will be necessary control at accessing the data at the system level. There is also a need to ensure that users can only alter the information they are legitimately authorized to change.

All organizations will want to gain confidence in their data integrity, and the key is to ensure that transactions across their systems are secure from tampering. The most common way hackers gain access to the data will be to obtain the necessary credentials of any user, but the best will be administrator access. With this access, the hacker will be able to initiate actions like deleting database records and suppressing any alert system (SMS/email) for such unauthorized access.

As compared to confidentiality protection, the protection of data integrity is beyond intentional breaches. The other countermeasure protection that can be put in place to protect data integrity will be access control and rigorous authentication to prevent authorized users from making unauthorized changes. Having digital signatures can also help ensure that transactions are authentic and the files are not modified or corrupted.

In a nutshell, effective integrity countermeasures must also protect against unintentional alteration. This alteration includes user errors or data loss as a result of system malfunction. Equally crucial to protecting data integrity are administrative controls such as separation of duties and training.

Availability

Availability is the third part of the CIA triad, and it helps to ensure that authorized parties can access the information they need. Regardless of how secure your data is, it will eventually fail the cybersecurity value if it takes a long time to retrieve the information when needed. To prevent the availability of data, the hacker will need to deny access to the data.

The most common way of denying access to the information in any organization will be using distributed denial of service attacks (DDoS). In today's world, you will hear news of high-profile websites being taken down by DDoS attacks. There are several other ways hackers can affect availability, and they include:

- Denial of service (DoS) and distributed denial of service (DDoS) attacks on servers.
- Ransomware attacks, which encrypt the data on targeted computers so that the authorized parties cannot use it. The goal of this attack is to compel the victim to pay a ransom to an attacker.
- Deliberately disrupting a server room's power supply to take those servers offline.

Most times, a company's information is more vulnerable to data availability threats than the other CIA triad components. To limit the damage that hard drives may suffer due to a server failure, natural disasters, or cyberattackers, it is crucial that you make regular off-site backups. Information is only valuable when the right individuals can access it at the right time. Some of the excellent information security measures for preventing threats to the availability of data include the following:

- Proper monitoring
- Off-site backups
- Disaster recovery
- Server clustering
- Failover
- Environmental controls
- Virtualization
- Continuity of operation
- Redundancy

Cybersecurity is one such niche within the field that offers plenty of exciting job opportunities for those who have the skills needed to carry out those duties. Like every other type of work, anyone can learn to become a cybersecurity expert with a basic intelligence level and plenty of *hard* work.

Takeaway: The internet is the enabler for productivity and also opens the gate to hacking. The increase in cybersecurity also leads to a rise in cyberattacks.

Chapter 2
Cyberattack and the Fishing Rod

Cybersecurity refers to the body of technologies, processes, and practices designed to *protect* networks, devices, programs, and data from attack, damage, or unauthorized access. There are several types of cybersecurity, and we shall be looking at the most common ones, but before then, we shall be considering some of the factors that motivate cyberattackers to hack into a website or application.

What Exactly Motivates Hackers?

Hacking is a serious threat and a source of concern for the entire information technology community based on the data showing how hackers have gone in recent times. Records show that there will be over seventy-five billion internet-connected devices by 2025, which also means that every one of those devices will be vulnerable to hackers.

As the number of these devices increases, so does the number of hackers, and they become more sophisticated. We have different types of hackers and various types of cyberattacks, and each one of these hackers has what motivates them to engage in hacking. Here are some of the most common reasons why people engage in hacking.

Cyber Hackers Engage in Hacking Just for Fun

Although cyber hackers may hack into a website just to make a few bucks and, in some cases, make good money, many cybercriminals actually engage in the act for the fun of it—just for a few laughs. It is interesting to note that about two-thirds of all cyberattacks on web applications are carried out just for fun.

A new Verizon report on over sixty-three thousand confirmed security issues—including web attacks, data breaches, and payment skimming in different industries such as mining and retail—confirmed this cybersecurity report. This is also one of the things cyberattack shares with the fishing game. Fishing is seen by many as a relaxing and social activity. Many people embrace it simply because of the promised relaxation and satisfaction they can enjoy while engaging in the game.

The fishing game offers people the added benefit of combining a serene landscape and productive waters' profit. Some cyber hackers declare that hacking into a secure system is one fun way of testing their wits and skills as well as that sense of adventure, just like the fishing game that leads many to the wild. They simply hack just because they desire to do it and have no intention of doing anything with the information except to have fun and gain some hacking experience.

Hacking for Theft

This is also another common factor that motivates cyber hackers. One of the things we hear regularly is the news of hackers who infiltrated an organization's database of social security or credit card information. The reason hackers want such information is that they can use it to steal all their victims' identities or generate duplicate credit cards.

When a hacker steals a person's identity, they can easily open up credit card accounts in the name of their victim and channel thousands of dollars or euros in charges. Hackers can enjoy a good return on the effort, time, and other resources they invested in hacking by targeting industries or small businesses with lax security rules.

Sometimes, hacking smaller shops are more lucrative for hackers than targeting large corporations with advanced security measures. But just like the fishing game where commercial fishers have a big fishing boat and net that enable them to launch their net and catch more fish, hackers would need to upgrade their tools before going after the big companies.

Espionage

Sometimes, cyberattackers engage in hacking for espionage purposes, and their main focus is often to get two types of information—manufacturing and political information. There are several cases where hackers leveraged government and corporate secrets for gain. For instance, in 2016, ThyssenKrupp, a German steel company, was not so lucky as they noticed that hackers had compromised trading secrets about their steel production as well as manufacturing plant design divisions. The company added that it would be impossible to estimate the extent of damage the attack caused considering the breach's nature.

Other hackers may also be motivated to carry out an espionage attack by gaining insight into a country's military or political intelligence. This was the case of the 2016 US presidential elections, which many believed was influenced by the activities of hackers who wormed their way to extremely sensitive data by spreading several phishing emails to employees at the DNC. They made efforts to influence the election with the data they stole in the attack.

Top 10 Most Common Types of Cyberattacks

- Denial of service (DoS) and distributed denial of service (DDoS) attacks
- Man in the middle (MitM) attack

- Phishing and spear-phishing attacks
- Drive-by attack
- Password attack
- SQL injection attack
- Cross-site scripting (XSS) attack
- Eavesdropping attack

The Fishing Rod

Traditional hacking aims to comprise the security of the settings of the IT system. With technology evolving, there are still basic methods hackers use to control the hacking game. Before the '90s, cyberattacks required a trained professional sitting behind the computer and wielding hacking tools to perform any attack.

Typically, such attacks are not carried out intentionally but just for the fame and fun of it. However, in today's environment, which has significantly connected almost everyone to the internet, hacking is done more intentionally and planned for some time.

Fishing is perhaps one of the most exciting and popular outdoor recreational activities, with over fifty-five million persons engaging in it in the United States alone. Although this number has dropped in recent years, this information underscores the popularity and value of fishing. But there is one critical virtue required to enjoy the fishing game, and that is *patience*!

A cyberattack is very similar to how a fishing game is played because both games require patience and luck. Skill and understanding of the game will result in victory as well. In a fishing game, an individual will undergo short training to understand the game and the equipment.

They would learn to pack the right equipment and select the perfect location where they expect a school of fishes to hang around their bait. This also applies to cybersecurity attacks. The trained individual also needs to understand both the game and their equipment. The fisherman uses a fishing rod, while the cyberattacker needs a computer and applications to launch an attack.

As earlier mentioned, a fisherman needs to select the right equipment and understand the game by finding the perfect spot to start. He needs to study the kind of fish he wants to catch and the area where it will be best to locate and capture the prey. This consideration is very critical, as it is the beginning of the fishing game, and if not done correctly, the game might get delayed, or they will not catch the prey. It is like finding a needle in the middle of the ocean; hackers will have the most challenging time locating their target.

Moreover, the prey's location and type will influence your decision on the specific fishing tool and bait you need to use. If the fisherman wants to go into a fishing game with a buddy, he will also need to consider if his buddy has the same knowledge and experience to catch the prey. If not, even when the prey appears, the tools and the fishermen will just watch as the prey escapes. And the ridiculous thing is that they will never catch the identified target with the fishing rod game. Instead, we will need a more sophisticated technique. Trying to catch a whale or a giant shark will never succeed with just a fishing rod.

This is similar to the cyberattacker who needs to understand the industry and find out more information about the company—the one that is the best and easier target. Of course, the returns at the end of the day must be profitable. Fishers will fish for a day or a few days, and one of the things that motivate them is that they will have enough fish to eat at the end of the fishing game. They could also use the trip as a great way to relax and stay close to nature.

On the other hand, cyberattackers hack a company to ensure they gain enough prizes in the game. The prize can be in monetary terms or just revenge. However, the most important consideration is if it is possible to crack the victim using any given method. A beginner cyberattacker is unlikely to take a multinational company (MNC) or government sector with a similar tool.

Using a fishing rod is the cheapest and most basic tool for the fishing game. Similarly, a cyberattacker using the fishing rod theory will be sitting behind his computer screen with the right hacking tools. The attacker can only reach out to one target/company at any point in time. Some experienced fishermen might be capable of handling two to three fishing rods simultaneously.

Just think of a father bringing his young kids to fishing and they are not focusing. The poor father has to manage all the fishing rods by himself. In the beginning, he might not be used to it; but over time, he will be able to handle this situation.

The maximum a more experienced hacker can stretch will also be two to three computers at one go. Cyberattackers using the fishing rod method rely on the cheapest method with limited resources and can only depend on themselves. In both situations, the biggest challenge will be when the hook gets one prey; they need to give all their focus and time to secure the catch. At this point, there would not be other bandwidth for them to focus on others. Once they lose focus, the prey will escape, and it will be tougher to have the same target tricked in the same area with the same bait.

The fisherman can be fishing in a team. This also applies to a cyberattacker who can attack as a member of a team. The typical scenario we will see in the fishing game using the fishing rod method will be a team of two. Such a buddy system will be scarier, as the two people pair up to source for their common prey. Having this buddy system will ensure a higher success rate, and not only will they back up each other in time of danger, but they will also always remind each other of their primary objectives. A team of two will make the fishing game swift, as they can make faster decisions with another party to double-check the game. The other buddy can wear a double hat and serve as a referee in the game.

However, we must understand that it will still be one person taking up one fishing rod at the end of the day and only focusing on one prey. Even if your buddy comes to your rescue to handle the other prey for you, when the rod catches its prey, they will also need to consider the opportunity cost of giving up the existing target or dropping it and focusing on the prey that just got caught.

It will all boil down to the engaged objective. But it is sad to say that most likely, the buddy will attend to his prey, as the fishing rod that caught the prey is customized and fine-tuned to capture the particular prey the buddy has in mind. The game can be a long-planned one, so why will he lose the opportunity to capture something he wants to save and assist yours?

The entire system requires strict adherence to the rules of engagement in this game. The success of the cyberattack or a fishing expedition depends on how well the fishermen adhere to the set rules.

Principle 1: The fishing rod can only catch one prey at a time.

The next step for the fishing rod game will be to enhance the tools—the fishing rod and the baits with the location and the kind of prey in mind and deciding on the buddy system. The fishing rod can be seen as the tools or applications on the cyberattacker's computer to crack the victim's end defense line.

When the bait is thrown into the water, the fisherman must be patient and wait for the movement underwater. A fisher might move the fishing rod a little to make the bait look alive from time to time. Taking up the bait still depends on the prey. Therefore, the fishing rod needs to be a more controllable item by the fisherman, and the bait should be strong enough to attract the prey.

The bait refers to the various techniques and methods used to attract the victim. The victim over here can be any individual and does not need to be in a corporate environment. Anyone working in the company is an individual who might or will access the internet after office hours or during the weekend. Besides using after-work and during weekend timing, this individual can also be a target of the cyberattacker when they access an unsecured wireless connection at a café or outside a public location.

The situation is worse for the victim, as companies engage in a policy known as bring your own device (BYOD) to allow their staff to bring their device. In such an organization, the staff will be using the same mobile phone for work (storing all work-related information on the phone). After office hours, the employee will also make use of the mobile phone for other personal needs. The employees will use their mobile phones to store and access work and personal email, which will put a big smile on the face of cyberattackers, as it increases their chances of success.

Besides mobile devices, the staff is also using the company's laptop to access personal information. Such personal data will be like personal email, access to a private online banking account, and running individual applications that are not scanned by the company's security team. These applications might be corrupted.

Some of these applications might carry malware, keylogger, and other hacking tools undetected by the staff. The situation will be worse if all the family members are using the same device as well. With such a multiuser environment, anyone who makes the wrong move will cause a cyberattack's success.

Principle 2: The chances of hacking will significantly increase with more users accessing a single device.

Chapter 3
The Hackers' Game and Strategies for Hacking

From the beginning of the cybersecurity game, the bait has always been the same. As technology evolves, hackers' techniques also improve with it, but the basics of the methods are still the same. Stealing passwords will be the easiest way for any cyberattacker. In the past, hackers can easily obtain people's passwords, as the system cannot handle complex password lists and they can easily guess passwords. System improvements have made it possible for various devices to take in more complex passwords. With secondary identification, the cyberattacker will have a more challenging time cracking the system.

Security Systems and the Five Elements of IAAA

All security systems function based on a principle known as IAAA. It is an abbreviation that stands for identification, authentication, authorization, and accountability. Here is a brief description of what each of the five elements means.

Identity

- First, a user would need to provide a name, username, employee number, ID number, SSN, and several others. A suitable security mechanism should identify any given identity. A user can define their identity by entering a username.

Authentication

This element refers to the password field on any website or log-in page. One way to prove your identity is by providing some form of password, PIN, and several others. Here are aspects of authentication you should know:

- *Type 1—something you know.* After logging in with your username, the next most common type of authorization will be using a password.
- *Type 2—something you have.* Users can also identify themselves using their identification token. This could be a smart card, passport, cookie on PC, and several others.
- *Type 3—something you are.* This will be physical identification, such as iris, fingerprint, or any other physical contact.
- *Type 4—somewhere you are.* It will be using your geographic location. Location-based authentication will be tricky, as a user needs to be in the set location and connect to the system.

- *Type 5—something you do.* It will be your keystroke and the pattern of accessing the system.

Once you successfully authenticate your identity, you are then provided with some rights, which takes us to the next element.

Authorization

- This has to do with what you are allowed to do and files or applications you can access. This requires the use of access control models, but how to implement the access control models depends on the company's goals. Some of the access models used by organizations include MAC, DAC, and RBAC.

Auditing/Accountability

- It is essential to trace every action on a website back to a particular individual.
- Also, we need to prove the specific actions someone took and when they did it under accountability or auditing. This is regarded as nonrepudiation.

For any access control to be considered strong, the system needs to fulfill two of the five elements. Even with such a strong access control in place, the bait is still strong enough to attract and catch its prey. There are other specially seasoned baits that the cyberattacker will use to ensure it appears attractive enough to catch the wanted prey. So let us go through the various strategies and baits that hackers use.

Phishing Email

The use of phishing emails has been widespread throughout the years. Have you ever imagined how many times people wish that they were the lucky one and that the phishing email they received is legitimate? If this were to be accurate, then there would have been hundreds or thousands of lottery winners. Using the phishing email method triggers the prey's emotion of greed. If anyone receives an email or link claiming that they won a prize, the target would be interested in clicking on the link.

There could be a prestudy on a particular prey before sending the link, or it could be a general bait that most ordinary people will likely need. The most common thought that comes to the prey's mind will be that "My computer already has an antivirus, and the window firewall is also turned on. Everything will be fine, and if the email is real, I will be the prize winner in no time." This kind of greedy mindset will motivate and encourage all prey to take up the bait and keep the fishing game going.

When the victim clicks on the link, their computer will be installed with a keylogger or other malicious applications running in the background. Moreover, this will grant the hackers access to monitor the victim's computer using the cyberattacker's control and command server. Once the control and command server picks up the credentials they are looking for, they will immediately launch the attack, or the installed malware will carry out the duties they are coded for.

The installed malicious application will be running in the background, and identifying it would be tough unless there is a monitoring or pattern detection solution. Such monitoring and pattern detection solutions will require most of the primary domains for cybersecurity to be set up. This requires time and money for the company to set up.

Ways to Detect Phishing Attack

Consider the following tips to help you detect phishing emails whenever you receive them in your inbox:

- Make sure you hover your mouse over hyperlinks in the email sent to you. This will enable you to see the real hyperlink address, and you can now carefully compare the link you received to the one you know. If you received an email from a coworker and are not sure if it is genuine, compare the hyperlink with your company's website address and cross-check the email address you received with your colleague's real email address.

- Ensure that you do not log in to a non-HTTP site and check the addresses to know if the letters have been switched or if there are typos.
- An email that contains a lot of grammar and spelling mistakes should raise the red flag immediately. Also bear in mind that scammers can sometimes clone existing emails that would appear very authentic and professional.
- Major companies and financial institutions, like banks, will not send you an email requesting that you confirm your password or provide security answers or card details. Report any email you receive that is asking you to disclose your personal information.
- Always view emails that are not relevant to you suspiciously. If you did not enter a competition, then you will not also win a prize. So when you get such an email in your inbox, do not respond to it.
- Some emails may try to convince you to send your personal information or money to claim a prize. Others may also apply the threat of closing your account if you fail to confirm your personal information. Ignore such emails and ensure you tag them as spam to stop them from entering your inbox.

Social Engineering

Social engineering will be another way of designing the bait to appeal precisely to a prey. It is also a typical and traditional way of getting information out from the target. Commonly, the hacker will create a modified email address of the victim's coworker or higher management. The hackers who disguise as internal staff often present themselves as members of the IT (information technology) department. They will be the best employees to win the victim's trust. The victim will be convinced that the fisherman posing as their company's IT person is legitimate and will reveal all the required information.

The information technology impostor can also send a link to the user and ask them to reset their password. But the information will be captured into another website, which will show the information. Moreover, they can access the system and hack into the administrative system with the information they got. Once they have access to the administrative system, they will be able to access the rooted server, which will be the beginning of a disaster.

Pretext calling will be another bait method that hackers use. With this strategy, hackers will create a scenario and obtain all the information they want. This is like a jigsaw puzzle; you fix up the mainframe, and now the only thing left is to obtain the parts and pieces of the puzzle. With the mainframe set up, you will know the victim to target, and this will narrow down the homework that the hacker needs to do. They will study the victim and work on specific questions and access methods to ensure the victims will be tricked. Once all the questions and access methods are established, they will play on trust to get closer to the victim and ultimately obtain all the information they want.

Pretext calling has a bigger success rate and is a quicker way to collect data than social engineering. This is because the method will establish the biggest trust link between the victim and the attacker.

Although physical breaches will have the highest success factor, it is also the hardest and most dangerous to execute. The only way to achieve this is to physically enter the premises of any company and pose as another party. This could mean duplicating an access card and gaining access by using their particulars.

To execute such a method, the hacker needs to carry out a thorough and in-depth study of the victim and their company, as they need to enter the premises and access and collect the data personally. These methods are the common and standard attacking tools that cyber hackers will use with their limited exposed resources and budget.

Other computer methods would assist a hacker in executing the mentioned methods but do not involve physical involvement. All the cyber hacker has to do is sit behind his computer screen and develop the best strategy for breaking into the company's defensive system. Such a fishing rod is also harmful and can take down a company.

Although using a fishing rod method can only target one company at a time, the key understanding is that if such a technique were to increase and power up their attacking strength, they could take down a massive network and have the ability to take over any company. Moreover, with artificial intelligence and machine learning, continuous hunting can be achieved, be it day or night. The hacking game will not stop but will grow stronger, bigger, and scarier.

Social Engineering and Social Media: A Hacker's Tool

There is no doubt that social media has helped many organizations to move their companies forward. It has become an excellent marketing tool for companies who desire to get their name out in front of potential customers and established customers without investing hundreds of thousands of dollars in advertising and marketing.

Remember, as businesses are increasing in their use of digital tools, cyber hackers also adopt new baits that would enable them to catch their prey. Most companies now turn to interns to help them create Facebook posts or regularly send out their tweets. In some cases, they hire a freelance social media expert to handle the job and ensure that the company's name remains in the public eye.

When they hire a freelance expert, all they will even pay will be a fraction of the cost of hiring a full-time employee. While social media has become an excellent marketing tool for companies, it could eventually be the most horrible nightmare for an organization, especially when hiring short-term interns or freelancers to handle their social media outreach.

So what is the risk associated with this means of communication? There is limited control on most social media platforms, and an organization can easily appear inexperienced and offensive, courtesy of an incorrect statement or a bad post. But what's even more challenging is when it is combined with social engineering.

Remember, one of the ways cyber hackers design their bait is via social engineering. What the fishermen would do with a company's social media posts is to obtain useful data regarding the employees and their duties. This is valuable information for hackers who are making efforts to engineer a hack socially.

Of course, negative comments regarding an employee can lead to a backlash, but an innocent commentary also poses a great deal of risk. A fisher will observe specific locations to identify the best fishing spot where to launch a net or cast a fishing rod. Location matters because a fisher cannot catch a big fish in a small river. He needs to go deeper, and this is where location comes in.

Similarly, cyberattackers understand that by reading the comments regarding employees, they will know who is the right person to be served with the bait. The right spot to get this information is on social media websites. Cyberattackers observe the comments and the things people say about the company—both good and bad. They are also monitoring those making the comments, and this is what they need to develop the right profiles that lead to targeted spear-phishing campaigns.

Cyberattackers can obtain specific details of the company's key decision-makers as well as other important personnel courtesy of the social media posts. The final result of their work is an extremely sophisticated socially engineered attack.

Social media provides hackers information about a person's preferences, which can make the illegal request appear more plausible. Here is an example of what a hacker could do with the information they got from social media. Remember, we earlier discussed the use of fear and greed as baits by hackers. Assuming the cyberattacker identifies a post on Facebook or Twitter from James Clinton who identified himself as the VP of IT technical support but has another name like Fred. When sending a phishing email, the hacker can write another employee or even call and say, "Fred requested that you confirm that your password was reset. Please send me your current password to enable me to validate it." It is easy for an employee to know that James is often called Fred because of how friends and colleagues tweeted to him on his social media account.

This form of personalized communication is a lethal combination when used alongside social engineering. Most hackers are aware of this strategy, and they exploit it frequently. Cyberattackers are also leveraging an organization's public perception, as it is presented on various social media websites. This is because these cybercriminals work with information, and they will collect as much information as they can get to explore ways to spoof customers.

Social media websites or applications have an immediacy effect that significantly affects the public and can instantly shift markets. People are easily encouraged to click on the link because it carries the kind of legitimacy that makes them trust other links. It is easy for people to carefully consider an email or link to a website from an unknown source.

But a victim's instinct to act or click immediately without thinking would be activated when a brand they trust (or what seems like a brand they trust) offers them a limited-time offer in their social feed. This will even get worse when another colleague or news source reposts the same link. This is a simple way to send the bait out both to the company's employees and even customers of the same company.

Companies should have a dual focus when it has to do with security for enterprise social media sites. An organization should examine its posts to shield its employees from socially engineered attacks and ensure that the organization's information is quite legitimate and not a bait or source of malicious activity.

This means that companies should ensure that the person or individuals in charge of managing the social media accounts are properly trained and informed about the company's security policies. The truth is that companies that already have a security policy in place and frequently discuss it with their employees are often better off than those who do not have one because they already acknowledge the risks.

The employees in such companies already know the risks, and this kind of awareness also leads to better security. Apart from good security policies, there is also a need for excellent security tools because of the volume and complexity of policies, regulations, and posts.

One of the best ways to handle social media security and compliance is by leveraging technology such as computerized systems that can easily scan and process information more consistently and faster than an outsourced service or department full of people. An excellent way to effectively address this security challenge is through a combination of technology and people.

Avoid Being a Victim of Phishing Attack

The emotions that hackers will use in equal measures to convince you to take action include the following:

- Curiosity
- Sadness
- Fear
- Guilt
- Excitement
- Anger

So to avoid being a victim of any of the social engineering attacks, such as phishing attacks, you need to be careful about what you do online.

It would be best if you cultivate the habit of doing certain things whenever you are online. Remember, cyberattackers would always try to make you act first before thinking later. So you need to slow down when you get a message that conveys a sense of urgency or makes use of extremely high-pressure sales tactics. The urgency of an email should never influence your careful review.

Make it a practice to be suspicious of all unsolicited messages even when the email appears to be from a top management person in your organization or from a financial institution you use. Research the facts using the search engine to get the company's actual website, phone directory, or other resources.

Never allow a link to control where you end up on the internet. Make it a habit to find any website yourself via the search engine and make sure you are taken to where you want to be. If you suspect an email, try hovering over links to see the real URL, but do not forget that a good fake URL would still mislead you.

Never download a link without knowing the sender personally or expecting a file from them. You will make the greatest mistake by downloading from any link you see in your inbox.

Reject all requests for help or any offer of help from companies or organizations because they do not really need to contact you to help them. Consider any offer to "help" do these things as a scam:

- Request to answer your question
- Restore credit scores
- Refinance a home
- Change your password
- Confirm your password

Delete any request for help from a charity organization that you have no relationship with. Always seek out credible charity organizations on your own to avoid being a victim of scams.

Denial of Service Attack

Denial of service attack (DoS) is an example of the kind of attack I talked about earlier—where the cyberattacker sits behind a computer screen and works out the attacking game. The DoS term can only be used in the fishing rod game, but with the upgrade of equipment and other tools, it will become a group of individuals attacking a system or a network of machines. The fishing rod will be a one-to-one attack in this denial service.

Denial of service attack is when a single cyberattacker attacks a single target. Only one computer will be working on a single internet network, and its goal is to flood the victim's internet with packets. It is an attack phase where an attacker sends many legitimate packets to the server and the server is unable to identify valid or invalid requests. With such a high volume of packets coming into the server or internet pipe, the system will be overwhelmed and unable to handle the volume of requests anymore—both valid and invalid requests.

Once the internet pipe is choked up, it will result in the victim being unable to access the internet, or the public will be unable to access their website server. Therefore, there will not be any established communication path, and the company will eventually be isolated from the internet.

DoS is an attack that involves overloading a company's computer resources, which makes the server inaccessible to others.

There are two common strategies for DoS attacks. It could be carried out as crashing services or flooding services. When it comes to flooding attacks, the system gets excess traffic, which causes the server to slow down significantly and finally stop. Among the widespread flood attacks are as follows:

- *ICMP flood.* This type of hacking strategy takes advantage of misconfigured network devices by pushing spoofed packets that end up pinging all the computers existing in any network that the hackers are targeting instead of just one particular machine. When this happens, the network will then also be triggered to amplify the traffic. Another common name for this kind of DoS attack is the ping of death or the smurf attack.
- *Buffer overflow attacks.* This is perhaps the most commonly used DoS attack, and it is designed to take advantage of bugs specific to some networks or applications.
- *SYN flood.* This kind of attack sends a request to connect to a server, but it still does not complete the handshake. It continues to send a request until it finally succeeds in saturating all the open ports with demands, leaving none of the open ports for the available and legitimate users to connect to.

The fisherman or cyberattacker will not stop there but will send even more requests and eventually overwhelm all the existing open ports on the network, which shuts the server down.

The second type of DoS attack is the crash attack, but this type of attack often happens less frequently. The cybercriminals will simply transmit bugs that will exploit identified vulnerabilities in any targeted victim, and this causes the entire system to crash.

The primary aim of DDoS attacks will be to deny any resource from accessing the website. Firstly, the downtime can be valuable to the organization, and not only will it disturb their reputation, but it also denies potential customers access to their website. Distributed denial of service (DDoS) attacks can lead to the unavailability of the organization's operation and eventually cause a company to wind up. Do note that other sources can also deny the service, like power outages or natural disasters such as floods.

Once DoS is achieved, a cyberattacker can move to the next step of action. DoS is commonly used in the business world to stop competitors from launching an online product or a marketing campaign. When your competitors find out that you are launching a marketing campaign that will gain a competitive advantage or a vital campaign that will increase your company's reputation, they will want to prevent your company from going through with their plan.

They will either hire cyberattackers directly sitting in-house or outsource the job to unknown attackers. Moreover, the easiest approach will be to deny the server so the campaign will not be launched successfully. This often weakens a company's reputation because whenever the public tries to use their website, they will not access it.

Service Unavailable

HTTP Error 503. The service is unavailable.

The same attack can also be used to deny competitors from entering a tender bid. Many tender requests happen on the internet, and it is the game of fastest fingers first. So the speed of the internet is very critical. Using the same method, your competitors can also deny your company access to the internet, so you will not be able to enter or access the internet to bid for the tender.

Why are there many fishing game victims over the years even though many people are having more knowledge and education about the cyber game? The answer is simply because the hacker leverages two aspects of our lives—greed and fear. Traditional cybersecurity protection will not protect the victim from these.

In the game of fear. If the staff received an email from management, they would be wondering, *What would happen to me if I fail to respond to the email? After all, there is a slim chance that the email is legitimate, and I will get into trouble if I do not follow up.* Moreover, the required information might be so vital that it will affect the company's performance or help them win a big project. The fear of losing their existing position will result in hackers winning this game.

The second one has to do with greed, and a good example is when you get a notification that you have won a million dollars and you need to pay a small fraction of the winning prize to secure this million. While you are reading the email and checking how to cash the million out, the hacking tool is already at work—retrieving all the information needed. It is just like when young people play a trading card game—the same mindset.

Whenever any player places the card of greed, they will be able to draw more cards and turn the table from losing to winning. So the card of greed is a happy tool in a card game, and it is also a comfortable tool in the hands of the cyber hacker.

Rule number 1: Cyberattackers love greed, trust, and fear.

How to Deal with DDoS Attacks

One of the best solutions to DDoS for a corporation will be to have a critical backup. Backup is always the key for any corporate data, and the best practice will be to conduct regular off-site backups. Having this in their company operation will allow corporations to control and limit the damage caused by hardware or natural disasters. And for an organization with highly critical information, having redundancy in its infrastructure will be most appropriate.

Off-site backup can be a hot, warm, or cold site. And regardless of the type of site the organization takes up, the off-site location data will be ready to restore the information when the primary data is affected. This is the best approach to reduce downtime when there is an issue on the primary site.

The key measures to ensure availability will include as follows:

- Implementing IPS systems and firewalls to prevent any DDoS attack
- Having an uninterrupted power supply (UPS) to back up your power supplies for the servers
- Backing up data to external drives or tape

When it comes to preventing DoS attacks, the general rule is that the earlier you can discover that such kind of criminal attack is in progress, the quicker you will be able to manage the situation correctly and contain the damage already done. Consider some of the options below to help you prevent this kind of attack.

You can seek professionals' help to identify the attacks. One of the tools that most companies use in defending themselves is anti-DDoS services, and you can also consider this option. The use of technology will help you easily identify legitimate spikes in your network traffic (especially during your business's peak periods) and a DDoS attack.

Consider getting in touch with your internet service provider. When you discover that your company is facing a serious threat to hackers, you need to quickly notify your internet service provider to help you ascertain whether your traffic can be rerouted. It would also be a smart idea for you to have a backup ISP in case of such attacks.

Some services help to disperse the massive DDoS traffic among a network of servers, and it would be a great idea for you to consider such services for your business or organization. Using this type of service would make a DoS attack ineffective.

Examine the black hole routing. Another option for internet service providers is to make use of black hole routing. This strategy can enable them to direct excessive traffic into a null route, which is regarded as a black hole. With this strategy, you will be able to prevent the targeted network or website from crashing as planned by the cybercriminals.

However, this strategy comes with one major challenge. When this is carried out, both the illegitimate and legitimate traffic will be rerouted just the same way.

You can also configure firewalls and routers. Your business or company can configure their firewalls and routers to reject bogus traffic. While doing this, always bear in mind that your firewalls and routers should be regularly updated to ensure that they get the latest security patches.

Make use of front-end hardware. Use of application front-end hardware that is added to the network just before the traffic gets to a server will give you the chance to analyze and even screen data packets. This type of hardware helps classify the data as a priority, regular or dangerous while entering the system.

With this tool, you will also block threatening data. It's important to note that businesses that operate on a small scale have a lower chance of becoming a target of DoS attacks than large-scale businesses that attract heavy traffic. However, it is still crucial that you take certain precautions to shield your business against such attacks.

Always ensure that your security software, applications, and operating system are frequently updated. One of the best ways to patch vulnerabilities that cyberattackers would want to exploit is through security updates. You can also consider making use of a router that has in-built DDoS protection. Choose a website hosting service that places a lot of emphasis on security.

What makes the difference between being prey and surviving the attacks of cybercriminals is taking precautions. This kind of precaution is often far more problematic for large organizations.

Man in the Middle (MITM) Attack

Another type of fishing rod cyberattack is the MITM attack, a general term for cybercriminals who position themselves in a conversation between an application and a user to either eavesdrop or impersonate one of the parties to make it seem like it is a regular exchange of information that is underway. So what exactly is the purpose of a MITM attack?

The perpetrator wants to steal personal information, like account details, log-in credentials, and credit card numbers. Most of the time, the targets of this kind of attack are e-commerce websites and SaaS businesses as well as other websites where users are required to log in before using the services. The information that the attackers get is used for different purposes, and one of them is to carry out unapproved funds transfers. Others include illicit password change and identity theft.

There are two distinct phases of the MITM execution, and they include the following:

- Interception
- Decryption

Interception

This is the first phase in the MITM attack progression, and it intercepts users' traffic via the attacker's network before it finally gets to its intended destination. Perhaps the most popular and simplest way to execute this is launching a passive attack where the attacker uses free and malicious Wi-Fi hot spots that you can find almost everywhere in public.

These Wi-Fi hot spots are often named to correspond to the location where they are provided. Also, these hot spots are not password-protected. The moment a prey makes the mistake of connecting to the hot spot, the attacker will immediately gain full visibility to any online data exchange. Some attackers desire to take a more active approach to their game of interception. For those in that category, they can choose to launch any of these attacks:

I. *ARP spoofing.* This involves a process of linking the MAC address of the attacker with the IP address of a legitimate user on a local area network using ARP messages. Consequently, the information that is transferred by the user to the host IP address will be sent to the attacker.

II. *IP spoofing.* In this case, the attacker will disguise himself as an application by modifying packet headers in an IP address. The result is that users who are making efforts to access a URL connected to the application are sent to the website of the attacker.

III. *DNS spoofing.* Another name for this kind of interception is DNS cache poisoning. This has to do with infiltrating a DNS server and making changes to the address record of a website. The result would send the users making efforts to access the website to the attacker's website by the altered DNS record.

Decryption

The second phase of a successful MITM interception is decryption. After the interception, any two-way SSL traffic has to be decrypted, and this should be achieved without alerting the application or user. To achieve this, cyberattackers can adopt any of these methods that I would share now.

I. *SSL hijacking.* This happens when an attacker sends a forged authentication key to an application and a user during a TCP handshake. This process eventually creates what seems to be a secure connection even when the man in the middle is controlling the entire session.

II. *HTTPS spoofing.* In this case, the attacker will send a phony certificate to the prey's browser after making the initial connection request to the secure website. The phony certificate holds a digital thumbprint linked to the compromised application that the browser will also validate based on an existing list of trusted websites. When done, the attacker will access any information entered by the victim before it will be passed to the application.

III. *SSL stripping.* By simply intercepting the TLS authentication that is sent from the application to the user, a hacker will downgrade an HTTPS connection to HTTP. The cybercriminal will then send an encrypted version of the application's website to the user and still maintain the secured session with the application. During this period, the entire session of a user will be visible to the hacker.

How Do You Prevent an MITM Attack?

To successfully block MITM attacks, a user would need to take several practical steps in addition to a combination of encryption and verification methods for applications. Users need to take the steps below:

a. Stay away from making use of Wi-Fi connections that are not protected with a password.
b. Always pay good attention to all browser notifications that report a site to be unsecured.
c. Always log out of a secure application when you are not making use of it.

d. As a rule of thumb, avoid using public networks you can find in hotels, restaurants, coffee shops, etc.) when engaging in sensitive transactions.

A smart option for website operators to mitigate spoofing attacks is to use a secure communication protocol (TLS and HTTPS) to encrypt and authenticate transmitted data robustly. By doing this, operators of a website will prevent site traffic interception and further block the decryption of sensitive data, like authentication tokens.

One of the best practices for most applications is to secure all the pages of their website by using SSL/TLS and not just the pages that need users to log in. This step would help decrease an attacker's chances of laying hands-on session cookies from any user browsing on a section of an unsecured website while being logged in.

Chapter 4
Powering Up the Game: Upgrading to the Fishing Net

When you look at what technology was in the '90s, you would agree that it is evolving rapidly, and more trained and experienced cyberattackers are getting involved each day. Just as technology is becoming very sophisticated, the cyber game is also changing its level. The tools and applications they use are getting more sophisticated and easier to manage. They are also able to target more prey at one go with less restriction. Like the fishing rod game, the fisherman needs to know the game's level-up tools and be familiar with them.

The more familiar they are with the tools and applications they use for their fishing game, the better they will be able to achieve their result. With technology evolving rapidly, the fishing game also needs to upgrade the tools and applications. The fishermen also need to be aware that they have to increase the tools to stay afloat of the game or be a part of the cyber game's revenue.

There is also an understanding that if all the experienced cyberattackers are using or upgrading themselves from using a fishing rod to the fishing net and you are still using a fishing rod, then you will lose out in the game.

This is because your competitors will catch your intended prey.

While you are still using a fishing rod to slowly catch one prey at a time and fine-tuning the method to increase the success rate, others who are using the fishing net have already got your target and are already trying their best to hack into their system. Even when you can successfully hook the prey, the prey may suffer from the previous attack from the fishing net.

The whole security system might have been enhanced already due to the previous attack. What is worse for a cyberattacker using a fishing rod is that the cyberattackers using the fishing net already carried or robbed all the items and information they want. Alternatively, they might have already left a back door for their future access so that they can come back to the same victim anytime.

However, the cyberattacker using a fishing rod typically has a lesser capability in terms of technology or might not be experienced compared to others. They can be the scapegoat for the cyberattacker using a fishing net. After the cyberattacker uses the fishing net to get all the information he wants, he will erase all traces and allow or guide the cyberattacker using a fishing rod into the victim's network. When the victim becomes aware of the attack and carries out all the checks, they will only end up tracing it back to the cyberattacker using a fishing rod.

In the fishing net game, the tools and applications will need to change. It is no longer going to be the fisherman holding on to the fishing rod. This time, the fisherman will be using a fishing net to catch the prey. What is slightly different from the cyber game is that the tools and applications might not change.

As these are virtual tools, they only need to fine-tune the coding or application to make the new tools suitable to continue or upgrade the fishing game. It might be cheaper virtually to upgrade the game, as they can improve and upgrade the application when they are behind the computer screen.

Alternatively, they can go around sourcing for a better application that is available in the market. They will have to weigh the differences between sourcing the applications externally and developing them internally. The time to develop the tools is also a key concern because they will be late in the game if they spend longer time to complete and continue to fish for prey.

In cyber hacking, timing is a critical factor to consider. If this element is loosely or not carefully considered, the whole battle can be lost, and others who are faster and better in the game will succeed in throwing a few punches at the prey. Moreover, hardware might not be required, though this depends on how the initial fishing rod tool is being set up.

If the fundamental is already established and well-thought-out initially, it will be a lot easier to upgrade the tools. What the cyberattacker can do with the existing tool (which is the same source code) is to efficiently use it to create another new tool to suit the fishing net's ability. In general, building and upgrading from the initial applications is still a more favorable option for cyberattackers.

They are also using this upgrade to fine-tune and readjust the methods they will be using for the new game. The new strategy can be thought up and drawn up to catch bigger and more preys. New logic will be defined and aligned to ensure higher chances of success. Also, a cloud solution can be considered as another resource to use and is a cheaper alternative. Cyberattackers have to weigh the resources and the processing power required for such attacks.

Principle 3: Upgrading the game ensures a higher rate of success with less effort and time.

Coming to another conclusion on the new tools, there is also a possibility that new hacking tools and applications need to be developed. If the initial fishing rod is purchased off the shelf for fishing rod game only and not suitable for other levels, the tools need to change. There is a higher chance that the fishing rod is developed based on a minimum scope and, therefore, unable to support the fishing net game. Planning and scaling up or down at any time will require a lot of input into the initial planning and setup.

They will need to think of how to manage their activities when there is a scale-up. How can computing power be put into a form that can be so rapidly deployed when there is a scale-down, and how can they save the existing computing power cost? If the new application and tools are known, there would be a new cost to purchase stuff.

Cyberattackers can assemble their hacking tools in different ways. First, they can buy off-the-shelf tools and enhance them to suit their needs, especially when they have a diverse target audience. Alternatively, they can develop new tools from scratch. Similar to upgrading the existing tools, cyberattackers need to take note of the timing. The longer the time, the more they will lose out.

The fishing net can provide a more straightforward game for the fisherman. In the fishing rod game, they need to know their prey and understand the bait. They need to create or mix or work out the best tricks that will attract the targeted prey. Typically, this will require prior engagement with the prey or in-depth research on the prey.

It is about finding out everything about the prey and deriving the best strategy for the hook. It also requires being a step ahead of the attack phase as well. A cyberattacker cannot just depend on the hook; they need to plan for the attack process as well. Once the victim is hooked, they will need to know the attack procedures required to hack into the account.

Some cyberattackers have no idea what to do next and how to do it after successfully hacking into an account. The timing they left behind while working out the attack strategy has allowed either another cyberattacker to come into the picture or the victim to spot the issue and enhance all the cybersecurity positions. This weakens the situation, and the attackers would need to refind their target. This is like finding the best methods to enter the game with the prey.

However, there is no need for the bait in a fishing net game, which makes the game simpler. As technology improves, the ways to catch these preys will be easier and smoother. Many things will be automatic, and hacking can happen around the clock with artificial intelligence and machine learning method.

While the cyberattacker is resting, the fishing net will be actively carrying out its duty. While the cyberattacker is sleeping at home—along the way from the fishing net—the fishing net is busy ensuring that all the prey close to the net will not escape its casting range. Even with this upgrade from a fishing rod to a fishing net, a fisherman can be stationed at the same spot.

What the fishing net does is that it will spread across a bigger and larger radar. However, unknowingly as mentioned, the fishing net will be able to catch unexpected preys that pass by. Location can be a challenge if the initial spot is small but leads to a big ocean because the fisherman will not be able to cast its net.

It will be a big challenge if you are fishing at a small opening leading to the big ocean and expecting the big fish to swim into the small spot. The initial planning already has a mistake. So upgrading the tools and applications gives space for cyberattackers to review their strategy and tools. If the location is an issue, hackers will need to find a better spot. With the upgrade in tools and applications, the tools can help locate the sweet spot for hacking.

Principle 4: Advanced hacking tools make the hacking process sweeter.

With the tools and location fine-tuned, next will be the prey that a cyberattacker expects in the net. There will be expected prey, which means a target that the cyberattacker already has in mind and is working on. These victims have been well-studied, and once they are hooked, the attack process that has already been drafted out will continue until the hacking is successful. There is prey classified as unexpected prey. They are victims who swim into the net even when it was not initially targeted at them.

In the cyberattack, it is like when you are targeting the financial industry and scanning across all financial industry IP but another company in another industry uses the same equipment and defensive system. So with the same setup, the scanning will be able to penetrate these companies as well. This unexpected prey might be part of the ecosystem from the financial industry.

The unexpected prey can be from the upper or lower part of the supply chain. Why such an ecosystem is targeted is because IP can be exposed accidentally. When a financial company's IT department is outsourced, the outsourced company does not care or is not cybertrained. The IP addresses and other information might be leaked out accidentally, putting the company in the same group during the scan using the fishing net method.

Another weakness is when a company outsources to too many companies. If any companies fail to do a proper job or create a loophole in the process and information is leaked out, the company will be swimming directly into the fishing net itself. By the time the company realizes what has happened, it would have become a disaster, and it will be very tough to trace who leaked the information. It is like water flowing into too many lakes—very tough to control.

Moreover, there is no primary water source, as each outsourced department will have its own guided water source. Outsourcing is the way out for a company that does not have any expertise. The primary risk of outsourcing is that the company will lose control and end up being controlled by the outsourcing vendor.

Therefore, a virtual CISO will be a more rational way to overcome this.

Today, most companies do not set aside a budget for cybersecurity, and cybersecurity is still part of the IT (information technology) cost center. Moreover, IT is a cost center for all companies. This explains why a company will not bother to hire a full-time cybersecurity team. The chief information security officer is the key person in the top management to answer all cybersecurity issues.

The person reports to the chief technology officer (CTO) and will offer advice on the company's strategy and operations. Such a high position will require the company to spend to maintain the position, especially when considering that anyone who occupies the position must have vast industry experience and provide answers to all cybersecurity issues.

Most organizations cannot afford a full-time position for this role, and therefore, they will engage a virtual CISO (vCISO). A virtual CISO will then serve as a security practitioner and provide insight into the hiring organization's security. The individual will be on demand and usually carries out the duty on a part-time or remote basis.

Although the vCISO works on an on-demand basis (since he holds the highest position in the company for cybersecurity), he will need to answer to media or anyone if there are issues or questions regarding cybersecurity.

Getting back to the fishing net strategy. The buddy system can still work in the fishing net game just as it did in the fishing rod game. The buddy first needs to know how to operate a fishing net and will have the same or common objectives to carry out the fishing game. As we all know, fishing using the fishing net will ensure that more prey is caught, and the action that should be carried out has to be more intense.

When the fishing net has prey, it will catch many targets at one go, and every victim in the net can mean success for the cyberattacker. With the larger scale of success, the buddy might not need to have their independent fishing net, but they can work in pairs for each fishing net.

The reason is simple: one fishing rod might take hours to catch one prey, which means that the return on investment (ROI) is low and time-consuming even if the fisherman operates three to four fishing rods at one go. The overall returns will be low as well, and over time, it might not be worth the effort, as there may be no single prey for days. But a fishing net is different; once you cast it with your buddy and make sure the spot you chose has fish in it, it will be able to net some fishes. The fishes might be small, but at least when you calculate the return on investment, there will be some profit.

Also, with the fishing net method, there is a chance to catch unexpected and bigger fish. You might even have the opportunity to fish other creatures, like crab and all. A try of different flavor is possible with the fishing net. Having such a buddy system will increase productivity and overall help to catch more prey.

Principle 5: Larger scale generates higher ROI.

Casting the fishing net will increase the number of targets (companies) to be caught in the net, and it creates a more difficult situation for the targets to escape. The caught prey will also have a lesser chance of escape from the game. In cyberattack, this is similar to upping the game in the fishing game. Now a fisher can upgrade his tools and with a slight increase in his investment and have a few hours of training.

The same trained individual can operate a more significant attack scale at one go with his improved hacking tool. He can be on the same fishing spot and with the same buddy. The buddy will need to be trained on the same tools, and with time, he will be able to handle the prey together with his buddy.

With such an update, the fisherman now uses a more powerful weapon that can catch more prey at one go instead of using the single-target prey tool called the fishing rod. Even though the game is being leveled up now, the objective is still the same. Cyberattackers' objectives will remain the same as the fishing rod game, which is to fully hack into any prey that falls into the hacking tools' casting range.

The explored methods of attack will remain the same. The net attack is an improved version of the fishing rod game. However, for cyberattackers, the whole game's fundamentals still fall back to the same strategy as the fishing rod—greed and fear! Phishing email and social engineering work the same over here, but with enhanced hunting methods that target a specified group. The net can be designed to focus on the financial industry or other industries.

Remember, the net will be able to catch other unexpected preys as well. After the upgrade, the hacking codes used are often created to broadcast to the same industry or a group of companies with the same defense pattern or use the same defense tools. Once this net can go through one company, similar companies with common attributes will be in the target as well.

When the common attributes are established, the cyberattacker can sell this logarithm on the dark web. It can be sold for money or for fame to gain more popularity. In short, if this customized net can capture one financial company, then all the other financial companies can be a target, especially if they are using the same defensive method.

Having such a fishing tool in the fishing game is still a waiting game for hackers. The reward now is more extensive, more unexpected, and less time-consuming. You can take this as the "same class" learning method. Imagine a course of professional defenders taught by the same master, read the same books, went through all the materials, and have the same experience. They're not exposed to any other ideas or methods. If one of the classmates goes insane and becomes a cyber hacker, he will be able to take down the rest of his classmates because he understands their layered defense.

Therefore, cybersecurity should not be any solution that you can purchase off the shelf. Instead, it should be designed by your internal or virtual CISO, and the procedure of immediate response should be crafted for that company only. Companies can have a similar approach, but no two companies should have precisely the same steps and tools.

Even for virtual CISO (vCISO), they should review a company's cyber infrastructure setup to be best suited for their operation and not just copy and paste the solution. Direct mapping will also mean that all the risks and vulnerabilities of the existing setup will be mapped over the new setup.

Rule 2: Cyberattackers love companies with the same defensive process.

The success rate is often very slim when all the attack strategies still stay the same and one computer attacks a network on the internet pipe. There is an upgraded version of the attack known as distributed denial of service attack (DDoS) in the fishing net game.

A distributed denial of service attack is where more than one computer floods one targeted internet. It uses a community of zombies or bots (affected devices) distributed around the whole world. In the game of fishing net, the prey caught in the trap can be converted into zombies. The affected zombies could be printers, smart devices, or any device that can store resources and connect to the internet.

In the current environment, over twenty billion smart devices are running around. So with the net set up, you might not be the final victim that the attackers want, but they would like you to be part of their army of bots. The net that was cast is equipped with specialized malware that can transform you into one of their armies. This can spread through a compromised website, email attachment, or internetwork.

Once a company in the net is affected, it will be turned into a bot/zombie as part of the attacking army. Now the affected devices are connected to the command and control servers, waiting for the centralized machine's order.

The standard command for these zombies will be the target and the attack method. The army of bots is called a botnet (hundreds or thousands of bots). Anytime the botnet wants to launch an attack, they will receive the instruction from the command and control servers with the target IP address, and they all will attack one time.

In the fishing rod game, the victim only needs to deal with one attacker, so they can easily overcome such attacks by merely having more resources than them. Moreover, it is a lot easier for them to handle and focus on resolving one attacker. While in the fishing net game with a DDoS attack, the victim has to deal with thousands of requests. It is now much harder for the victim to withstand the attacks as it grows from DoS to DDoS.

Chapter 5
The Current Game: Fishing Boat

With the evolving technology and how rapidly it changes, cyberattackers no longer need to sit behind a computer screen to control the entire cyberattack cycle. The same goes for cyber defenders; they no longer need to stare at the screen because they can control the situation with artificial intelligence and machine learning. Now it means that the cyber game can happen and be executed anywhere, anytime, and by anyone.

No one needs to physically control the game, and all are done with the click of a few buttons. With the increase in the number of smart devices and IoT (Internet of Things), the cyberattacker does not require a laptop or computer desktop to manage or organize an attack. Attackers can organize and execute every attack from a mobile phone.

Anywhere

When someone is surfing the internet in a café using their access point, they can be a hacker or a victim. A hacker will be listening to any important packets behind their laptop or mobile device. It can be easily carried out using a program to crack any password once the victim falls into this trap. As a victim, you innocently log in to the access point and surf your personal information.

These packets can be captured when the hacker is on the same access points, making things easy. The worst part is that with such a heavy hacking tool, your firewall and virtual private network (VPN) will not be able to handle it but will eventually give way. So in a virtual environment, everyone can be the target anywhere. This can also happen overseas when you are on holiday mode and your awareness is low.

Anytime

Just by using logarithms and with computing power, hackers can attack at any time. With machine learning taking over, the hacking process can still take place in the background while the cyberattackers are resting or sleeping. However, there are still cyberattackers who form groups and prefer the manual way of hacking.

Typically, they will form a team, and each will attack round the clock in shifts. Without geolocation restriction, IP addresses can be redirected, and you can be hopping IP every second. The team can be formed across different countries. Without any limits, the other guy across the globe would take over if one person goes to sleep.

Since the computer does not need rest, they can farm these victims every second without any disruption. Also, an online community can be easily formed and disbanded once its objectives are met. There are hacking teams created to hack with specific goals in mind, and once the goals are achieved, the team will be disbanded and the teammates will start looking out for a new community to join.

Anyone

With the growth of the dark web and online forums, anyone can be the target unknowingly. If you offended a cyber hacker, they might display your identity on the dark web and challenge others to hack you. Moreover, the IP address is all that is required to conduct the hacking process. Your home access point is the most common access for hackers.

Generally at home, no one will have a firewall and other protections, so it purely depends on the password protection. Password protection is easy to hack into. Moreover, with the cyber game of using a fishing net, anyone can easily get hooked into the situation. The other common way is when you are in public and using your mobile phone.

When your hot spot or Bluetooth is turned on, you are opening up your access for cyberattackers. A simple layer of the password will be an easy task for cyberattackers. Once the cyberattacker manages to get into your mobile device, they can retrieve all your data and use them to either hack deeper into your environment or pick someone from your background to work on.

Principle 6: Cyberattacks can happen to anyone anytime and anywhere.

The most exciting and scary part of the cyber game is when the cyber environment is upgraded to a fishing boat. The most direct characteristic of a fishing boat is that it can travel across to a more profound and future part of the game. The ability to enter into the deeper ocean and travel to the future is an essential element, as these locations are where the big and lucrative fish swim in.

The reward of moving into the ocean is unbelievable, and the final aim of all experienced hackers is to arrive at the big sea. Now the cyber game is the same; the hacking game can hunt more preys around the clock. It will remove all the restrictions that the fishing rod and net have and fulfill all cyberattacks' features—anytime, anywhere, and anyone.

The fishing boat still holds the fishing net tool but has now added one more platform for the fisherman. If the fisherman or the team of hackers has more resource power in finance, they can add the boat motor. Large fish are not allowed to swim in where the fishing rod and the fishing net hunt.

It could be that large fish cannot swim into the small area or the small area is unable to attract them to swim into it. In terms of the cyber game, if they can catch a large fish, they will gain a high reputation, and it is well worth a lot more in terms of dollars and cents. Large fish have better defense and more extensive computing power, so it will be a lot tougher to take them down.

Therefore, a fishing boat can travel into the deep ocean for a bigger catch. Moreover, because there is a platform and they are not restricted by human movement, they can brave the storm. The weather conditions that limit what the fishing rod and fishing net can do have been removed.

With a bigger boat and better equipment, the cyber game can work on bigger fish round the clock, so location is no longer tied or restricted in the cyberattack game. In the game of the fishing boat, the fisherman can still travel alone. The fisherman can adequately equip the vessel with a boat motor and install the gear and other equipment to handle the fishing net. Having such a setup allows the fisherman to travel alone.

Alternatively, the fisherman can work out the fishing experience with a buddy for better support. The buddy can look out and share responsibility, which might be overlooked by the fisherman. This is just like in the cyberattacking game where the other buddy will be around to ensure the team is not being traced and all traces are erased or covered up.

In case when they require any support or countercheck, the buddy can back up. Still, the buddy or the team must be able to handle and be familiar with all the fishing boat tools. If there is a team, they must have their own assigned roles and responsibilities. Handling a bigger fishing boat will require more teamwork from the crew members, and the boat captain plays a critical role. In this aspect, a big fishing boat is controlled by a more organized team, and the team usually is led by an experienced captain.

Now we understand that fishing boats can overcome many restrictions like location, timing, and technology. The bigger and more sophisticated equipment the fishing boat is having, the bigger crew it will need and the more experienced professionals it will require. The fishing boat can even upgrade to a fishing harpoon to handle even more powerful preys.

With all the equipment and motors installed, the fishing game also needs to increase its budget. Usually, cyberattackers equipped with such technology could be a bigger company or government.

In the market, there are a lot of smaller fishing boats. A smaller fishing boat will be equipped with slightly simpler tools. These simple tools can be built up fast, created, and driven by anyone. Even a traditional fisherman can quickly learn the right skills required.

So what is the industry situation of the cyber game in the version of the fishing boat? In today's environment, the fishing boat's net will be bigger and cover a larger area. Once the fishing boat net is cast, hundreds to thousands can be their prey. If the fishing boat is installed with a bigger machine and runs automatically—keeping and launching of the fishing net—it can catch a bigger prey in its trap. If the target is too strong and has a strong defense, the fishing harpoon will be activated to ensure the prey will be hunted down.

In the cyber environment, once the prey encounters this hunter, they will be hacked easily. The attacker will possess strong, paid, and well-established hacking tools handled by very experienced professionals. Therefore, the prey will not be able to escape once they are caught inside the net, and regardless of how complex their system may be, they will be able to take it down.

The ocean does not in any way limit the reach of these fishing boats. They can even fly or travel on the ground. They can be present anywhere and at any time. When prey is the target of a few fishing boats or part of the unexpected victim, the victim can be hooked or trapped in multiple nets at one go. The resources for casting the net are ready, and the cyberattackers can cast the net at any time.

For example, John's company unknowingly is trapped under one net; and due to their weak positioning, another fishing boat came or passed by. The other fishing boat cast its net in the area, and John's company is also trapped in it. Now the company is trapped in two fishing nets, and the two attackers are trying their best to crack into the company's network. Unfortunately, the third and fourth boats came. Assuming John's company has the best defensive and monitoring cybersecurity solution, can their system handle up to hundreds or thousands of such fishing boats? Remember the DDoS that happened during the fishing net event. If one fishing boat issues a DDoS command to John's company, they might be able to handle it. But if hundreds of fishing boats issue the same command at one go, the large bandwidth will also deny the company bandwidth.

In the cyber world now, cyber hackers can be born at any time and anywhere. With the free online course, which anyone can access 24-7, and with so many online video channels, anyone can learn about cyber hacking. Universities are now offering many such cybersecurity courses and allow more and more exposure to cyber risk.

Therefore, everyone is talking about cybersecurity, and it is in the interests of many people. More media channels are also advertising and broadcasting information on cyber defense and attacks. With such awareness, the cyber game can increase the risk of an attack. The victim might be having a strong defense, but an amateur can manage to do a hairline crack in the system.

The first few hundred fishing boats may not break the strong defense, but the 999 fishing boats may eventually crack it. Besides, the next few hundred fishing boats can gain access and work out their goal.

Understanding the Game

The cybersecurity game can be broken down into three elements: anyone, anytime, and anywhere.

Anyone

In the fishing rod game, anyone can easily take up the fishing rod and join the fishing game. The same goes for the cyber game; once someone attends a few hacking course days, they can become a hacker. Upgrading to the fishing net requires a more experienced and mature hacker. Then moving up to the fishing boat would require a license and more resources. The resources will be human and financial resources that can only be supported by a company or country.

Anytime

Whether a cyber hacker wants to sleep or not, there will be times that they would be unable to fish in the fishing rod and fishing net game. The preys they are looking for might be sleeping or swimming in another area. There will be a time for the cyber game when corporations are doing their security patches or reviewing their system.

Cyberattackers will not want to be present during this timing, as the risk of being noticed will be higher. While in the fishing boat, if they fail to fish in a given area, they will move to another area rapidly. Also, with more resources on board, the fishing boat will not feel any restriction at all.

Anywhere

As a fisherman using a fishing rod, the hacker would need to stay near the shore because there is no other way to go deeper into the sea unless they are using a boat. To jump into the fishing boat game, the fisherman needs to pour out extensive funding to upgrade. The same applies to the fishing net; the net can be cast near the shore. Moreover, when we are near the shore, the bigger fish will not be swimming there, but the fishing boat can overcome the geolocation concerns because of its speed (boat motor) and toughness.

Due to globalization, it is easy to form a community, and the buddy system can be easily tied up. With such a team and community, the support will be there, and this allows the team to endure the initial hacking phases, and they are also more experienced as a team. It will be tougher to take them down. With the increase in cyberattacks cases, companies need to be aware of the situation and prepare themselves.

Afterword

There are few notes about the information communication technologies evolution.

If someone from the IT professionals started to write a manuscript about contemporary internet technology trends or even if you prepared a short presentation last night, you would hardly be on the top news of the industrial development level. Such an intensive, even dramatical, new industrial philosophy with a digital machine shakes the world population similar to an earthquake.

Primarily, the information communication technologies (ICT) for nations communication worked as international partnership tools. At the beginning of a new millennium, many of us must remember the frustration related to the magic expectations about digit 2000 forecasting many systems collapse. Many doubts and risks existed with a very intensive dot.com development involving both professional and public society toward the unknown and ambiguous world of new life's investment and behavior.

Nations should have a closer look at information and communication technologies that allow doing business and cooperating internationally. So what is the central root of the problem?

The problem is that the ICT is new for the world's nations. The ICT power surrounds you on the earth, sea, land, and underground as well as in space. There are eleven thousand satellites around the Earth, which are fulfilling more than eight billion devices on our hands today. By 2025, their amount will have reached thirty thousand satellite objects. However, the cyberattacks are so massive that they have profoundly affected economies and societies, which brought to a large number of victims of ICT.

So to contemplate this virtual life, similar to the discovery of the space trail, we have to carefully study not only the NASA platform or Cape Canaveral in Florida but also the evolution of technologies that permit us to achieve such progress. That is why the human race should understand that their passion for stars and space should be in balance with their passion for the earth.

The absence of the internet will cost economies more than $50 billion daily and will damage devices linked to essential services, such as transport or health care. For instance, in 2003, one of the fastest worms in history, the Slammer/Sapphire worm, infected over seventy-five thousand devices in ten minutes and eleven million within twenty-four hours, showing us how vulnerable we are in the digital world. Today, our dependence is rising exponentially, meaning that we should be better prepared for risk mitigation and its management.

In the previous decades before the new millennium, significant threats were raised from nuclear technologies, a large number of atomic bombs and military resources, as well as political behavior. Now it is about the digitalization potential and ecological threats. There are four components of threats to the use of ICT: interference in the internal affairs of the state, cyberterrorism, the military-political aspect, and cybercrime. However, it becomes a severe weapon in opening the global digital war, such as the country's presidential election, terrorism, and illegal immigration worldwide.

Unfortunately, there are no standard rules for global cybersecurity yet. Most of the countries are creating defending systems through different bilateral agreements to ensure their nation and country's safety. The UN secretary recently stated that the next crisis would emerge from a malicious attack carried out by pros, madmen, and punks. The question is whether most countries can withstand such damage from intensive and costly attacks.

There is no international containment mechanism. The world is moving away from international law. There are 158,000 international treaties and memorandums for today's agreement. However, nobody knows when article 51 of the UN, the right to self-defense, comes into force and how to use this right. The second reason—as stated by the US, UK, and Australia—is the pursuit of a proactive policy against cyberattacks. Countries' cyber demarcation in cyberspace needs to be fixed via international agreements and consensus. Those are the major threats and challenges today.

The exponential growth in the creation and adoption of the ICT industry over the past two decades has accelerated progress and shortened the periods of industrial revolutions, minimizing a century to a decade. As some high-tech executives have said—for example, Bill Gates and others—"In 2020 we made a jump into the reality of 2030."

Artificial intelligence is susceptible to cyberattacks just like an ordinary person. A hacker can change the program, encrypt or decrypt the messages, and easily threaten the survival of humanity. Edward Snowden revealed the principle of surveillance of all, a global spy network tracking. At the UN level, it is recognized that society does not know how to manage cyber technologies but only begins to formulate rules based on common sense principles.

The paradox of information power is that the more developed the information and communication technological potential is, the greater is the fear of vulnerability. At an earlier time, after the WWII and during Cold War, it was the enormous number of nuclear bomb resources accumulated on the earth that could destroy the world and its population. Horrible nuclear weapons were only used twice, in Hiroshima and Nagasaki, completely destroying cities.

However, today, cyberattacks have been used repeatedly for a long time and are the ones harming the world. Due to the high speed of cyberattacks and the damage that it causes, a cyberwar can arise more rapidly than a nuclear one. Cyberweapons are faster and more devastating than any other rifle. The episodes are not carried out directly through intermediaries from any country. The military of most of the developed countries creates cyber teams and electronic bombs, in addition to cyber propaganda, fake news, and so on. Their verification is complicated but possible. Australia has announced having a cyberattack verify technology; and Singapore, during its cyber weeks, demonstrates trends and makes predictions in the ICT industry.

Some experts talk about cyber sustainability from the point of view of environmental sustainability or economic sustainability. By using transparency measures, confidence-building measures, international law cyber norms, etc., they attain this sustainability. Moreover, the help of military intelligence has helped the US, UK, and other states to diminish the number of cyberattacks in one way or another and lower the threshold of war conflict. So the stability dialogue is working on the security problem.

Using different possible options, nations should learn to navigate to the future. Fast Internet of Things, Industrial Internet of Things, and the creation of a digital twin are the possible options.

There is a cascade of pressure on the economy, politics, and human life. The transition to a postindustrial world leads, of course, to a digital war for market superiority in all areas of the world.

Here is why it is really important to orient the youngest generation to study this field, support their ideas, and help them to become more competitive and effective. And for the future I said as a senior and hire professional I understood search requirements when I went through supervising several PhD students in IIoT, IoT, and cybersecurity.

Final Thoughts

Cyberattacks have consistently occurred over the years, and this is one reason why no organization or business should overlook this aspect of their business. The rate at which cyberattacks arise across the globe is quite alarming. Unfortunately, no business that has an online presence is in any way immune to such attacks; and the physical, financial, and legal implications of a cyberattack on any corporation or small business can be devastating.

One of the best ways to create a stronger defense is to discover your company's weak spots. When you identify your company's weak points, you will focus better on dealing with the potential threats of a hack. Another way to put up a strong defense is to understand the hackers' game, which is the focus of this book. When you learn about various tools, baits, and hacking applications, you will be better positioned to prevent attacks.

The primary focus of this book is to expose the average hacker by describing their activities like a fishing game. It will help all IT and cybersecurity professionals to make the right security decisions for their organization.

References

Agrawal, G. 2019. "'IAAAA' ... Five Elements of AAA Service." Retrieved on January 1, 2021, from https://mrcissp.com/2019/01/16/iaaaa-five-elements-of-aaa-service/.

Anand, P. 2014. "Most Hackers Do It for Fun, Not Profit." Retrieved on December 30, 2020, from https://www.marketwatch.com/story/hackers-do-it-for-fun-not-profit-2014-04-23.

Chivers, K. 2020. "What Is a Man-in-the-Middle Attack?" Retrieved on January 3, 2020, from https://us.norton.com/internetsecurity-wifi-what-is-a-man-in-the-middle-attack.html.

CloudFlare. "What Is a Denial-of-Service (DoS) Attack?" Retrieved on January 3, 2020, from https://www.cloudflare.com/learning/ddos/glossary/denial-of-service/.

Imperva. (nd). "Man in the Middle (MITM) Attack. What Is MITM Attack?" Retrieved on January 3, 2020, from https://www.imperva.com/learn/application-security/man-in-the-middle-attack-mitm/.

Kaspersky. (nd). "What Is Social Engineering?" Retrieved on January 3, 2020, from https://www.kaspersky.com/resource-center/definitions/what-is-social-engineering.

Milkovich, D. 2020. "15 Alarming Cyber Security Facts and Stats." Retrieved on January 1, 2021, from https://www.cybintsolutions.com/cyber-security-facts-stats/.

Paloalto. (nd). "What Is a Denial-of-Service Attack (DoS)?" Retrieved on January 3, 2020, from https://www.paloaltonetworks.com/cyberpedia/what-is-a-denial-of-service-attack-dos#:~:text=A%20Denial%2Dof%2DService%20(,information%20that%20triggers%20a%20crash.

Pedersen, T. 2017. "CISSP—IAAA (Identification and Authentication, Authorization and Accountability)." Retrieved on December 31, 2020, from https://thorteaches.com/cissp-iaaa/.

Poremba, S. 2015. "Why Hackers Love Companies Who Use Social Media." Retrieved on January 3, 2020, from https://www.forbes.com/sites/sungardas/2015/02/24/why-hackers-love-companies-who-use-social-media/?sh=612ff02071a9.

Smart Eye Technology. (nd). "Confidentiality, Integrity, & Availability: Basics of Information Security." Retrieved on December 31, 2020, from https://smarteyetechnology.com/confidentiality

-integrity-availability-basics-of-information-security/.

Tyson, J. 2019). "What Is IAAA?" Retrieved on January 1, 2021, from https://blog.jamestyson.co.uk/what-is-iaaa.

Webroot. (nd). "What Is Social Engineering? Examples & Prevention Tips." Retrieved on January 3, 2020, from https://www.webroot.com/us/en/resources/tips-articles/what-is-social-engineering.

Weisman, S. 2020. "What Are Denial of Service (DoS) Attacks? DoS Attacks Explained." Retrieved on January 3, 2020, from https://us.norton.com/internetsecurity-emerging-threats-dos-attacks-explained.html.

About the Authors

Dr. Tan Kian Hua, the core author, a renowned cyber expert, is an emerging cybersecurity advisor whose expertise in data and information security helped various companies to grow their businesses and increase profitability dramatically.

He led organizations in looking for ways to expand their reach and revenues and to stay competitive. The companies he advised include SMBs (small-medium businesses) from businesses with two-employee staff strength to large multinational companies (MNC) of six-thousand staff strength.

During his six years with an MNC, he was chosen as a young leader to attend their leadership program. He spearheaded a team to build from scratch a first world-class defense cybersecurity infrastructure and rectified a cyberattack within one day—the market average is twenty-eight days.

Tan Kian Hua is currently the principal consultant in RFiD industry and the virtual chief information security officer (vCISO) for numerous companies.

He holds multiple professional certifications related to data privacy and cybersecurity:

- CIPP/US (Certified Information Privacy Professional/United States)
- CIPM (Certified Information Privacy Manager)
- CISM (Certified Information Security Manager)
- CISA (Certified Information Systems Auditor)
- CDPSE (Certified Data Privacy Solutions Engineer)
- CEH (Certified Ethical Hacker)
- ABDP (Advanced Big Data Professional)

He is passionate about creating awareness on the importance of cybersecurity in all companies and continues to ensure a first-class standard for maintaining cybersecurity procedures.

Dr. Vladimir Biruk is a co-owner and managing director of the nongovernment, noncommercial organization the Institute of Education Development in the phere of Governing and Communications "CAPITAL."

Although the beginning of his professional and scientific activity was based on technical specialties, the connection with information and communication technologies has been accompanying Dr. Vladimir Biruk for the full fifty years. With professional competence in mechanical engineering and as a qualified researcher, his activity is associated with developing many technical systems and industries: nuclear energy, oil and gas and woodworking industries, banking and financial sector, and diagnostics of complex technical and communication mechanisms. His job required collecting and analyzing a large amount of data, automation, inventions, and innovative proposals for various scientific and industrial complexes. The introduction of a control system and automation of processes created new devices accompanied by designs and active international cooperation.

The first remarkable achievement of the team research of SRI "HidroProject," Moscow, was under the direction of doctors of science A. P. Kirilov, V. B. Nikolaev, L. and K.

Luksha, in the field of nuclear energy and atomic power plant construction. The results of long-term tests and calculations of the strength characteristics of elements in a complex stress were included in the state standards of technical safety and control of the new-generation nuclear power plants under Rosatom and RusHydro.

The second significant achievement in information and communication technologies is transforming the scientific and industrial complex of the Soviet Union (newly independent states) into international integrated technological processes. He worked as a senior expert of the Association for the Development of Information Technologies, under the leadership of academician A. P. Velikhov, created by the USSR's government. ICT projects in the field of space, environmental monitoring, and unification of education were carried out, including the transformation of the Institute of Cybernetics of the Russian Academy of Sciences into the Program Systems Institute named after A. K. Ailamazyan. Dr. Ailamazyan was his teacher and manager at the DIT Association, and the institute is a leading Russian research institution in the realm of information technologies.

The third successful project is related to the development of banking and financial institutions in Russia (Central Bank of Russia), Belarus (National Bank), and other Eurasia countries. Dr. Vladimir Biruk took part in creating eight stock exchanges, several commercial banks, and the banking system. He was the initiator and leader of the interbank currency exchange. The company was originally set up for the shareholders, who put modest money in the amount of $185, then management multiplied its capital by one hundred thousand times within three and a half years of operations. The peculiarity of projects in creating innovative complexes was based on information technology, including trading, settlement information systems, and arbitration. That is a notable feature of this project, which required cryptographic support and special information and computer security conditions.

Over the past two decades, work has been carried out for international organizations such as the World Bank (doing business projects from 2002 to 2021), the European Commission, technical assistance projects for Central Eastern Europe (TACIS project), and Know-How Fund (UK).

Education activities are as follows: qualified as a Certified Management Consultant in IMC USA since 2009 and in the Republic of Belarus (PO BOMAC), which complies with ISO/IEC 17024 standards and ICMCI principles, and professor at the American online LIGS University and leading universities in Belarus, Russia (RANEPA).

Index

A

artificial intelligence, 20, 30, 35, 41
auditing/accountability, 17
authentication, 9, 16
authorization, 9, 16
availability, 7, 10, 25

B

bait, 14, 16–18, 20–21, 29–30
black hole routing, 25
buddy system, 14, 31–32, 39
BYOD (bring your own device), 15

C

CIA triad, 7–8, 10
confidentiality, 7–9
crash attack, 23
cyberattackers, 11–17, 20–25, 27–33, 35–38
cyberattacks, 6, 10–14, 30, 32, 36, 40–42
   most common types of, 12
cyber game, 6, 24, 28, 35–38
cyber hackers, 8, 11, 19–20, 24, 33, 36, 38
cybersecurity, 6–11, 31, 33, 38
cybersecurity game, 16, 38

D

data encryption, 9
data integrity, 9
DDoS (distributed denial of service) attacks, 10, 23–26, 33–34, 38
   how to deal with, 24
decryption, 27
DoS (denial of service) attacks, 10, 22–23, 25

F

fear, 24, 32
firewalls and routers, 25
fishing, 11–14, 30
fishing boat, 35–39
fishing game, 11–14, 17, 28, 31–33, 37–38
fishing net, 28–33, 36–39
   game, 28–31, 33–34, 38
   method, 31–32
fishing rod, 11–14, 19–20, 23, 28–30, 32, 36, 38–39
   game, 13–14, 22, 28–29, 31–32, 34, 38
   method, 14, 20
flood attacks, 23
front-end hardware, 25

G

greed, 17, 24, 32

H

hacker attack, 6–7
hackers, 9–13, 16, 18–23, 27, 30, 33, 35–36, 38–39, 41
    motivations
        for espionage, 12
        for fun, 11
        for theft, 11
hacking, 10–12, 15–16, 30, 35

I

IAAA (identification, authentication, authorization, and accountability), 16
ICT (information communication technologies), 40–41
identity, 16
information security, 7
integrity, 7–9
interception, 26–27
internet, 6–7, 10, 12, 24, 33, 40
IoT (Internet of Things), 5–6, 35

M

machine learning, 20, 35
MITM (man in the middle) attack, 26–27
    how to prevent, 27

O

off-site backups, 10, 24–25
outsourcing, 31

P

patience, 13
phishing attacks, avoiding, 21–22
phishing emails, 17, 21, 32
    how to detect, 17–18
pretext calling, 18
prey
    expected, 30
    unexpected, 30–31

R

ransomware attacks, 10

S

social engineering, 18, 20, 22, 32
social media, 20–21

T

timing, 29–30

V

vCISO (virtual CISO), 31, 33

www.ingramcontent.com/pod-product-compliance
Lightning Source LLC
Chambersburg PA
CBHW030841180526
45163CB00004B/1415